BRAD CANNOT ADD

WRITTEN AND ILLUSTRATED BY BERNIE SCHWARTZ

Brad Cannot Add
Copyright © 2024 by Bernie Schwartz

ISBN: 979-8895311103 (sc)
ISBN: 979-8895311110 (e)

All rights reserved. No part of this publication may be reproduced, distributed, or transmitted in any form or by any means, including photocopying, recording, or other electronic or mechanical methods, without the prior written permission of the publisher and/or the author, except in the case of brief quotations embodied in critical reviews and other noncommercial uses permitted by copyright law.

The views expressed in this book are solely those of the author and do not necessarily reflect the views of the publisher, and the publisher hereby disclaims any responsibility for them.

Writers' Branding
(877) 608-6550
www.writersbranding.com
media@writersbranding.com

My brother, Brad, was sad.

Brad showed our Dad his report card.

Brad had received a D in math.

Dad asked Brad, "Why do you have a D in math?"

"I cannot add," said Brad.

"Is it bad that I cannot add?" asked Brad.

Dad replied, "No it is not that bad but you need help to add."

"Let me call my friend Chad. He is a math teacher," said Dad.

So, Dad called Chad.

"Chad can you help my son, Brad, learn to add?" asked Dad.

Chad said, "I will come over tomorrow to help the lad add."

The next day Chad arrived at our house, to help Brad add.

Brad looked a tad sad when he saw Chad at the table.

Chad said, "Come over here Brad and let us talk about how to add."

Brad sat down next to Chad
and told him why he could not add.

So, Chad pulled out a pencil and a pad to help teach Brad to add.

Chad wrote on the pad some numbers.

"Now Brad let's see how to add these numbers," said Chad.

Brad looked at the numbers and then looked at Chad.

"I do not know how to add these numbers," said Brad.

Chad looked at the lad
and then showed him how to add.

After a few hours, Brad was adding the numbers.

Brad was so happy that he could finally add.

They all shouted,

"BRAD CAN FINALLY ADD!"

THE END

From The Author….

This book is intended to teach children how to read by using rhymes with "ad". The book also shows children not to be afraid to get help with their school work or any other problem they may have. You can learn or solve any problem when you get help.

This is the third book I have written. "Days of the Week with Farmer Clay and Fay" and "Fat Rat Lost His Hat" are my first two books.
As I write and illustrate them I am inspired to continue writing children's books.

God, Family, and Friends have supported me in this effort and I hope you and your family enjoy my books.

www.ingramcontent.com/pod-product-compliance
Lightning Source LLC
LaVergne TN
LVHW070443070526
838199LV00036B/689